TRUFFLES

Pierre-Jean Pebeyre

TRUFFLES

KEN HOM
PIERRE-JEAN PÉBEYRE

Photographs by Jean-Pierre Gabriel

serindia
contemporary

TAGLIATELLES
AUX TRUFFES
TRUFFES BROSSÉES
SAUCE AU FOIE GRAS
TRUFFÉE
FOIE GRAS
PEBEYRE
Truffes
SAUCE AUX TRUFFES

CONTENTS

PREFACE

I first discovered the joys of French cooking about forty years ago. It was love at first sight, and first taste: the people, the land, the culture and especially the delightful and glorious French cuisine. But as I tasted, and sampled, and feasted my way through the colourful food markets and restaurants of the various towns and cities, the one ingredient that truly captured my imagination was the truffle. Ever since my introduction to the black truffle of Perigord, I have been experimenting with that unique aromatic fungus in cooking. And what an enviable job it has been!

As I read all I could on this mysterious and wonderful ingredient, I constantly came across the name Pébeyre. It was so identified with truffles that I was determined one day to meet Alain Pébeyre, whose face graced countless

food magazines and articles. As with so many other happy events in one's life, serendipity played the key role. As I developed my skills as a chef and sought always to extend the reach of my abilities, I met Eric Brand in the mid-1980s. He was then Food and Beverage Manager at that Hong Kong landmark, the famed Peninsula Hotel, and was kind enough to introduce me to Jacques Pébeyre, the son of Alain. We became quick and fast friends and soon I became an extended member of his family. Jacques' son, Pierre-Jean Pébeyre, became the brother I never had. Through this remarkable family, I was given the opportunity to learn more about the mysteries of the black truffle.

During more than two decades of deepening friendship with the Pébeyre family, I also absorbed the essence of the black truffle. In discussing the subject of the truffle, I was impressed by Pierre-Jean's extensive knowledge of all aspects of the truffle and its history. Although this knowledge was prodigious, it was his passion and love for the subject that amazed me and that I felt it had to be shared with the rest of the world. Hence this book.

In this book, Pierre-Jean and I share with you our approach to using truffles that is mindful of French traditions but remains simple, direct, even lighthearted. I incorporate various techniques and ideas from my Chinese culinary background. Is it authentic? Well, it is the meeting of East and West, which is what my collaboration with Pierre-Jean is all about.

My hope is that you will use this book as a handy and helpful guide to this unusual ingredient and that you will discover, as I have, how extraordinary black truffles can be.

Ken Hom

Left: Tinned truffles sold by Pébeyre. The label was drawn by Pierre Pébeyre, the founder of the company in 1897.

Below: Pierre-Jean Pébeyre and Ken Hom.

What Do We Know About the Truffle?

Pierre-Jean Pébeyre

THE PARADOX OF THE TRUFFLE

The truffle is a food surrounded by mystery. Its price and rarity and the extraordinary aroma that permeates it—all these give it a special status. And yet these mysteries are not always to be found where they might be expected. You might even say that there is a baffling lack of mystery on this subject. What I mean by this is that certain beliefs, certain aspects of a mythology that one might expect to find concerning the truffle, quite simply do not exist. While being an object of veneration and respect, the truffle has not given rise to a mystique as complex or profound as those of other precious substances, such as tobacco, coffee, certain perfumes and rare spices. However we look at it or study the truffle—from the point of view of its history or legend, of botany or cuisine—it is, above all, a paradoxical fungus. Historically, scientists and writers, mycologists and poets have sought to understand the phenomenon. They have never found themselves in complete agreement. The truffle has never been fully defined and the question "what is a truffle?" has never been answered in a completely satisfactory way. The truffle is a bizarre product in every way. Here is an organism that some have hesitated to classify as a fungus, and others as belonging to the plant kingdom. According to dictionaries and treatises, it is a "very delicious and fragrant plant" (*Dictionnaire de l'Académie*), "a type of unformed fungus" (Boiste), "a sort of stemless and rootless

Page showing different types of truffle from "Die Hypogaeen Deutschlands—Truffeln und der Diesen Verwandten Organismen" by the German botanist, Hesse, published in 1891.

Truffle orchard in South-West France in winter, the truffle harvesting season.

fungus" (Noël) or simply a "distinctive tuber" (Moynier), which is the description that best suits it. "We have long sought to know", writes Moynier in his 1836 *Traité de la truffe*, "whether it should be classified among the plants or among the minerals [...]. Nevertheless, [scholars] have pronounced in favour of the plant kingdom, with the comment that the truffle is born, lives and dies [...] it was classified among animalized plants."

Scientifically, we could not be more undecided. Whatever it is, this curious animalized vegetable mineral grows entirely underground, with no light and little oxygen, which is unusual enough. Also, the truffle is black; if not entirely black inside where it is crisscrossed by white veins, at least black on the outside, which is atypical of a living plant. In terms of growing season, it is also unlike everything else because it only grows at the height of

winter and its quality declines at a time when other plants are preparing to bloom. Neither does it follow the example of fungi, which love the cool dampness of autumn or spring mists.

To summarize, with the truffle, we have all the ingredients for the creation of a fantastical legend and a status as a magical fungus. Yet, the truffle is neither of these. It has the impudence to thrive in lands of legends and fairy tales, with rich peasant traditions, while rarely being associated with any myth or assigned a supernatural status. This strange, completely black thing that grows underground in winter; this thing that does not grow everywhere, whose presence is accompanied by circles of parched grass on the surface; this rare and capricious tuber that does not seem to obey any rule, that we have never succeeded in cultivating; this strange thing, finally, whose wonderful aroma has sour accents and has been imbued with aphrodisiac qualities (this alone would be sufficient to produce a fairy tale) has never given rise to any witchcraft tradition. In its long history there has been no connection with fairies, wizards or demons. We have preferred to concentrate on it aromatic qualities, and everyone is largely satisfied with this. This is a curious ethnological phenomenon; it is as if the mystery of the truffle ought to remain intact, to such an extent that, contrary to habits, even the popular imagination has decided to leave it in peace.

Basket of black truffles at the end of a day's hunting.

Truffle landscape near Cahors, South-West France.

The truffle: virtual or real?

You might say to me: "That's amazing, because there's no shortage of folklore surrounding the truffle. The black truffle is a legend in its own right; what's more there are no shortage of truffle brotherhoods. It even has a patron saint, Saint Antoine, and truffle masses are held in Richerenches, near Valréas, in the Drôme. Some of these must date back to the Middle Ages." Well! No. The truffle masses have been performed for no more than the past thirty years. It was two brokers from the Papal enclave, Jean-Marie Valayer and Raymond Mure, who had the idea to remove two or three truffles from each basket at the Saturday market. They collected them together in a shoebox and arranged with the curate to celebrate a mass on the theme of truffles—the effect of which was to amply fill the church. It was simply an attempt to elegantly rejuvenate the parish, but the formula was successful. The "tradition" was brought back quite recently, and has become part of the modern, artificial, mythology of the truffle. Since it has become institutionalized, the truffle mass has become a fashionable event; Richerenches parishioners are unable to get into the church on these momentous days.

The folklorization of the truffle is thus a recent phenomenon. It could not be left in its original simple state, it was too much in demand in the twentieth century. So everything was done to fill a void that had not sought to be filled. Under the pretext that it is a rare, expensive and mysterious product, there has been a desire to surround the truffle in a pseudo culture it did not possess in previous centuries. People have used an adulterated idea of *terroir* and a trashy version of the Middle Ages in support of this. This folklorization reminds me of fake rustic furniture sold in superstores on the outskirts of towns. Imitation furniture in poorly weathered walnut or a cherry that is too dark, with none of the delightful irregularities possessed by country furniture. The world of the truffle has a schizophrenic tendency because it is a double world; a virtual world of the truffle exists alongside the real one. The truffle does not need to be legitimized; it has no need of legends. However, those who have seized on legends to make it a fairground attraction are not those who know it best. Virtual truffle aficionados exist vicariously through truffles. They influence government and the media. Their truffle culture, centred on a love of high society life, is coincidentally the one that makes the most noise.

A recently unearthed black winter truffle.

The future of the truffle

This mindset has not helped professionals and public authorities to react quickly enough to arrest the decline in truffle production over the past half a century. Indeed, during this time, truffle production has withered, and it runs the risk of vanishing altogether. No one in the virtual world of the truffle has truly devoted himself to finding a remedy or given serious thought to it, just as we have watched Venice sink into the mud without finding a way to save it. The real issues have not been tackled: the future of the truffle, its potential cultivation, etc. If the truffle disappears one day, perhaps we will create a commemorative encyclopaedia, *Memories of the Truffle*, and nostalgic publications; meanwhile the foundations of the trade are ensured a little while yet. But this vegetal diamond will become extinct without us giving it the opportunity to survive. The fewer truffles there are, the more they are talked about.

The truffle is not an abstraction, not a sort of god ensconced on a velvet cushion before which we prostrate ourselves; it is a vegetable and a condiment. It is a magnificent vegetable and a unique condiment, but a vegetable and a condiment nonetheless. We should not lose site of its cuisine when talking about the truffle. Currently, there are enough people who keep trotting out a past that never existed to ensure that this culinary treasure is confined to a ghetto as a luxury product and above all one that is tied to classical French cuisine. And yet, currently the majority of truffles are consumed outside of France. It is time to be creative, to explore and to acquaint the modern world with the truffle and all its culinary richness.

Chinese wonton ravioli with truffles. See page 112 for the recipe for this delicious dish.

This is where my friend Ken Hom comes in. With his expertise and knowledge of his own tradition and his enthusiasm as a globetrotter, he has enriched the truffle's repertoire, giving it a new lease on life. Thanks to his very eastern knowledge of flavour harmonies, he highlights the nobility of the fungus rather than making it sacred. Having lived in Quercy for the last few years, he is a true truffle aficionado. In this book, he brings together his own traditions with ours, mixing his recipes and our family recipes to make a harmonious whole, a wonderful balance of flavours and traditions.

I would love other taste innovators, from you with your own cooking to the most celebrated chefs, to follow suit and open-mindedly explore the true potential of the truffle.

THE TRUFFLE: AN ENIGMA

The fruit of an underground exchange

The truffle is the fruit of an underground mycelium that lives in partnership with the roots of trees, generally oaks, but also, among others, pines, hazels and hornbeams. The mycelium of the truffle provides the tree with mineral substances that increase the tree's capacity to absorb nutrients, while the tree provides carbon products thanks to photosynthesis. It is a mutually beneficial exchange and not, as might be believed, a parasitic relationship. The point of contact between the tree and the mycelium is in the root system, within a particular organ, the mycorrhiza. The link is thus not a direct one. Since the truffle has a life that is totally autonomous, a tree alone is not sufficient to produce truffles. The mycelium has to grow in a particular soil and climate: generally aerated, clay-limestone soil with a pH of around 8; the climate must be temperate—cold in winter, warm and wet in spring and warm and stormy in summer.

These requirements explain why truffles are not found everywhere. Only the northern Mediterranean currently seems to favour their growth. Provence has always had a strong truffle tradition, at least when it comes to truffle production and commercialization. In the nineteenth century, substantial planting and reforestation efforts were made. Important Provençal figures, like Martin Ravel and Rousseau, put their minds to the development of the cultivation of this product, and large trading houses (for example Badouin and Rousseau) grew up there.

The largest markets took place there, in Carpentras and Montagnac. But the truffle had difficulty finding its way into Provençal cuisine. There are few traditional recipes from this region that use the ingredient. Perhaps this is because this cuisine is largely oriented towards the sea and makes great use of olives, which compete with rather than complement truffles. This is the difference between the south-east and the south-west of France, and Perigord in particular, where the truffle has long been part of the culinary tradition and has become a selling point for other products: for example, pâté de foie gras and fattened chickens with truffles.

This constant association between truffles and the region's cuisine explains the generic name "Perigord truffle" for the *Tuber melanosporum*.

A black truffle buried in the ground, freshly discovered by a truffle hunter.

TRUFFLES FROM FRANCE AND ELSEWERE

Truffle producing regions

Truffles can only be found in three European countries:

- in France, across the southern rim of the Massif Central and mainly in two regions: the South-West (in the Lot and Dordogne) and the South-East (in the Rhône valley) south of Valence, in particular in the Drôme, Vaucluse, Gard and the Alpes-de-Haute-Provence;
- in Spain, where truffles are harvested in Calania and in a quite large area between Valencia and Madrid around Teruel, Morella and in the Guadalajara region;
- and in Italy, where they are found in Piedmont and in the centre: Marchem, Umbria and the Abruzzo peninsula.

New production regions have appeared in the southern hemisphere, firstly in Australia. But wild truffles were not previously found there, and these are planted truffle orchards. The story began around thirty years ago when Yann Hall, a New Zealand researcher, attempted to produce black truffles; he obtained his first tuber at the beginning of the 1980s. The experiment was repeated in Tasmania, and then on the Australian mainland. Many plantations appeared in the regions around Sydney and Melbourne and above all south of Perth. This introduction benefited from the crucial work of a mycologist, Nick Malajuck, who carried out the efficient and effective cultivation of newly planted truffle orchards. Australian truffles, which are of a comparable quality to their European counterparts, have the advantage, for us, of arriving in market out of season. Australian production, which is currently growing, could eventually mean fresh truffles available in winter and summer in both hemispheres.

Another country, Chile, has recently become interested in the truffle. The first truffle orchards planted there are still too young to produce truffles, but in a few years, Chilean production should rival that of Australia.

The emergence of new producing countries represents a true revolution. In the aging world of the European truffle it is a welcome breath of fresh air.

The village of Morella in Spain, at the heart of a truffle-producing region.

Different types of truffles

When mature, Perigord truffles look like irregularly shaped tubers, of variable size and firm consistency, covered with a grainy skin called the *peridium*. This peridium on the *Tuber melanosporum* is covered with small black polygonal diamond-shaped lumps. The inside of the truffle is dense, dark grey or black in colour, crisscrossed by a network of fine pale veins. As the truffle matures its veins become finer and disappear and its texture loses its firmness. The whole fungus gives off a strong, powerful and aromatic odour, which seems to contain the essence of the soil. Black truffle growth begins in spring and they take seven to ten months to reach full maturity. The truffle's principal enemies are winter frosts and excessive heat or drought in summer. After harvesting, truffles can be kept fresh for up to two weeks on a bed of earth or sand.

Another type, the winter truffle (*Tuber brumale*), is harvested in much smaller quantities than the *Tuber melanosporum*; even though its production is growing, it still only represents ten to twelve percent of the volume of Perigord truffles harvested each year.

The two types are very similar in size, shape and colour; they grow in the same conditions and are harvested at the same time of year, but the veins in the *Tuber brumale* are thicker and wider spaced. In addition, its skin comes away more easily on brushing, which gives it a somewhat pallid look. Its musky flavour and aroma are less strong than those of the Perigord truffle.

Italy is where the famous white Piedmont truffles (*Tuber magnatum*) are harvested. They have a very different look and feel to other types. Their skins are smooth and pale and light brown in colour. When fresh, they give

Freshly washed black truffles.

off a very strong alliaceous aroma, which is totally lost when they are cooked or sterilized. They should therefore be eaten raw, generally grated or finely chopped. White Piedmont truffles often grow at the feet of poplars, near streams and at the bottom of ditches. They are harvested in autumn, when the earth is damp. They are very high-quality truffles, which obtain prices up to three times those for Perigord truffles. But they should not be confused with white summer truffles, which are of little interest. From a commercial point of view, internationally, Piedmont truffles are much more strongly marketed than black truffles; they are better known in wealthy countries, such that, in the United States for example, when they talk about "truffles" they nearly always mean white truffles.

The Chinese truffle is a recent discovery. Its botanical name, *Tuber indicum*, makes reference to India because the first ones were discovered in the northeast Himalayas. This type was first described and named by Cooke and Massie in 1892. It was only a century later, in 1994, that the variety was rediscovered by Japanese and French mushroom growers in the south of Sichuan and north of Yunnan. This Chinese truffle resembles its sister, the European black truffle, but without having its flavour or aromatic strength. Nevertheless, this fungus can be pleasing in cooking, at a much more affordable price.

Page from "Fungi hypogaei—Histoire et monographie des champignons hypogés" by the Tulasne brothers, published in 1851.

The quality of Chinese truffles is not consistent throughout their production region. The best varieties come in particular from the province of Shangri-la in Yunnan, which produces good quality truffles. The production, which was exclusively exported to Europe, Japan and the United States, is increasingly frequently consumed in China itself. Harvesting methods are different

Truffle labels from the Alphonse Pécoul truffle house in Carpentras, France.

from those in Europe, where harvesting calls on animals' senses of smell. Chinese truffles, which are found at high altitudes, between 2000 and 3000 metres, are harvested by hand by digging randomly under shrubs. This method of harvesting, as might be suspected, is very destructive. It is to be hoped that the Chinese will adopt more precise harvesting methods.

The summer truffle (*Tuber aestivum*), finally, is mostly produced in Italy and to a lesser extent in France and Spain. It does not have the right to the appellation "truffle" and can only be sold as a "white summer truffle". Despite the fact that it is harvested in summer, it grows in the same conditions as the Perigord truffle.

Certainly it looks similar in size and skin colour, but it is clearly distinguished by its yellowish flesh and blandness. This resemblance has often been at the root of frauds, but a microscopic examination of spores shows the very clear differences between the two varieties.

From time to time people have sought to "invent" new types of truffles, which are generally defined by the regions from where they are unearthed. Hence the existence of a "Meuse truffle" and a "Burgundy truffle" with brown flesh and heavy white veining. In reality, the Meuse truffle is a *Tuber mesentericum*, and the Burgundy truffle is a *Tuber uncinatum*, which is just a variety of the summer truffle, *Tuber aestivum*. Areas that favour truffles are generally dry, arid and uncultivated land, with a soil pH of between 7.5 and 8, i.e., limestone or sandy. Truffles from sandy soils have a more regular shape than others.

The rediscovery of the truffle

Truffles were eaten as long ago as the time of the ancient Romans. But their truffles were not the same as ours; they were *terfez*, truffles imported from various regions around the Mediterranean, especially from Libya. Light brown, these tubers did not have much flavour of their own, but acted as veritable flavour sponges, absorbing the taste of the dishes they were cooked with. What's more, they are still eaten in this way in North Africa and especially in Jewish cuisine.

The Romans may have come across black truffles from conquered Transalpine Gaul or from Umbria. More specifically, they may have delighted in Piedmont white truffles, which could be found just below their feet, but they preferred *terfez*. In truth, Roman cuisine was very rich in strong flavours. Their favourite condiment, *garum*, was a fermented fish sauce a little like Vietnamese *nouc mam* and other Asian fish sauces. A recipe can also be found in Apicius for *garum* with wine for truffles: pepper, lovage, coriander, rue, *garum*, honey, wine and a little oil—a union of strong flavours as I am sure you would appreciate. Indeed, the Romans were fond of spices in extravagant quantities; the flavour of true truffles would not have enjoyed the place it merits in such a context. The same could be said of the Middle Ages, which was the age of spices par excellence. Marinades, garlic dressings, anchovies, roasted bread with plenty of cinnamon, nutmeg, pepper, cloves, saffron, strong herbs like sage, mint, hyssop and rue, etc.—this era had its fill of strong and overwhelming flavours. But the truffle does not go well with other strong flavours.

To appreciate it you must not make it jealous; make sure you give it some space. This was not the time for this sort of cuisine. Even though it was hardly ever used in medieval festive cuisine, the truffle was nevertheless known about. The humanist agronomist Olivier de Serres for example, described the potato, the new vegetable from the Americas, as a "truffle". And indeed, in Europe, the common names for potatoes were often confused with those of the fungus (*tartufo, tartifle*, etc.).

In the sixteenth century, in France the wind began to change in gastronomy. Gradually, heavy and spiced sauces were abandoned in favour of fresher produce, vegetables, salads, finer cooking and less dressed-up products. Little by little, cooking with oriental spices gave way to the gastronomy of the Renaissance, seasoned with local spices, which included the truffle. It could now proffer the best of itself.

It was in the seventeenth century that it made its entry into European cuisines, and it could be found pretty much everywhere as the popularity of spices declined.

The golden age of the truffle

The black truffle gained its place in cooking in the seventeenth and eighteenth centuries. Although peasants had been eating them previously, it began to feature more frequently on the menus of the wealthy. The issue of storing and transporting truffles arose, since fresh truffles can only be preserved for a few days. Truffle trading companies were created in Paris and Lyon, sending brokers to the production areas to buy truffles, package them in barrels of brine and transport them to the large centres of consumption. Needless to say, their quality suffered greatly from such treatment. Until the beginning of the nineteenth century, the trade was carried out by traders from Lyon in particular. Every year they wintered in Paris between November and April to sell their merchandise, enjoying a quasi-monopoly position. The truffles were delivered to them every day from the Dauphiné and Provence via Lyon where they were packed. Before leaving Paris, the traders sold the last of the season's truffles conserved in oil or lard, delivered from Lyon in barrels of ten to one hundred kilos. Then they disappeared until the following winter.

François Appert's invention of the "canning" sterilization process ("*appertisation*" in French) and the advent of the railways radically changed the situation. While it is true that the culinary prestige of the truffle reached its apogee in the eighteenth century, it was the nineteenth century that was the true era of the truffle. It was, indeed, its greatest era of production and consumption. Canning enabled truffles to be preserved correctly, without destroying their flavour, and the railways considerably reduced the time it took to transport the tubers, ensuring regular supply and optimal freshness. "The truffle trade businesses in the Perigord were to those of Dauphiné–Provence what a dwarf is to a giant", wrote Moynier in 1836. However, with the help of technical progress, the regions of Perigord and Quercy, previously disadvantaged by the Lyonese brokers' monopoly, could finally grow and maximize their potential. At the local level, specialist truffle houses were born.

A late nineteenth-century print depicting women cleaning truffles by hand, taken from L'Illustration, 1880.

Throughout France, truffles found their way into restaurants and fine delicatessens, where huge plates of truffles sat enthroned as an essential decoration between pâtés de foie gras from Strasbourg and fine wines. Gourmets from all regions, particularly those from Paris, praised them to the skies and began to include them in abundance in their dishes: it was the time when recipes in cookbooks and in the ladies' press were prepared to begin with: "Have two pounds of good truffles ...". Literary gastronomes, like Fulbert-Dumonteil, devoted odes to them and pages of pretentious but enjoyable lyricism. This was the golden age of the truffle.

A food lover checks whether a fattened chicken with truffles has a good truffle smell. Taken from an 1828 publication: "A Manual for Truffle Enthusiasts, ..."

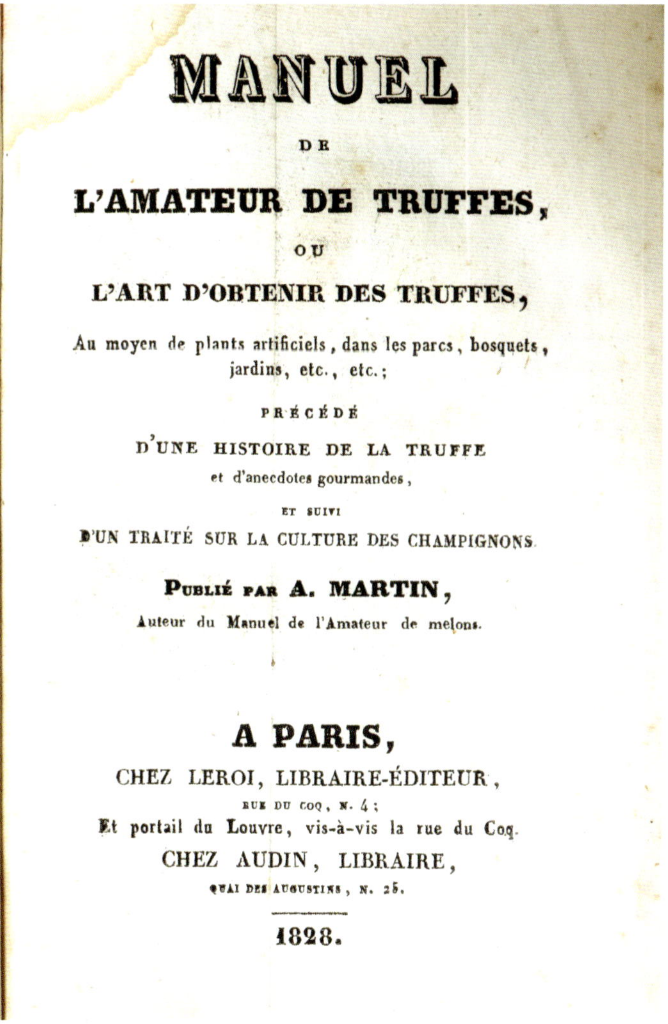

MANUEL

DE

L'AMATEUR DE TRUFFES,

OU

L'ART D'OBTENIR DES TRUFFES,

Au moyen de plants artificiels, dans les parcs, bosquets, jardins, etc., etc.;

PRÉCÉDÉ

D'UNE HISTOIRE DE LA TRUFFE

et d'anecdotes gourmandes,

ET SUIVI

D'UN TRAITÉ SUR LA CULTURE DES CHAMPIGNONS.

PUBLIÉ PAR A. MARTIN,

Auteur du Manuel de l'Amateur de melons.

A PARIS,

CHEZ LEROI, LIBRAIRE-ÉDITEUR,

RUE DU COQ, N. 4;

Et portail du Louvre, vis-à-vis la rue du Coq.

CHEZ AUDIN, LIBRAIRE,

QUAI DES AUGUSTINS, N. 25.

1828.

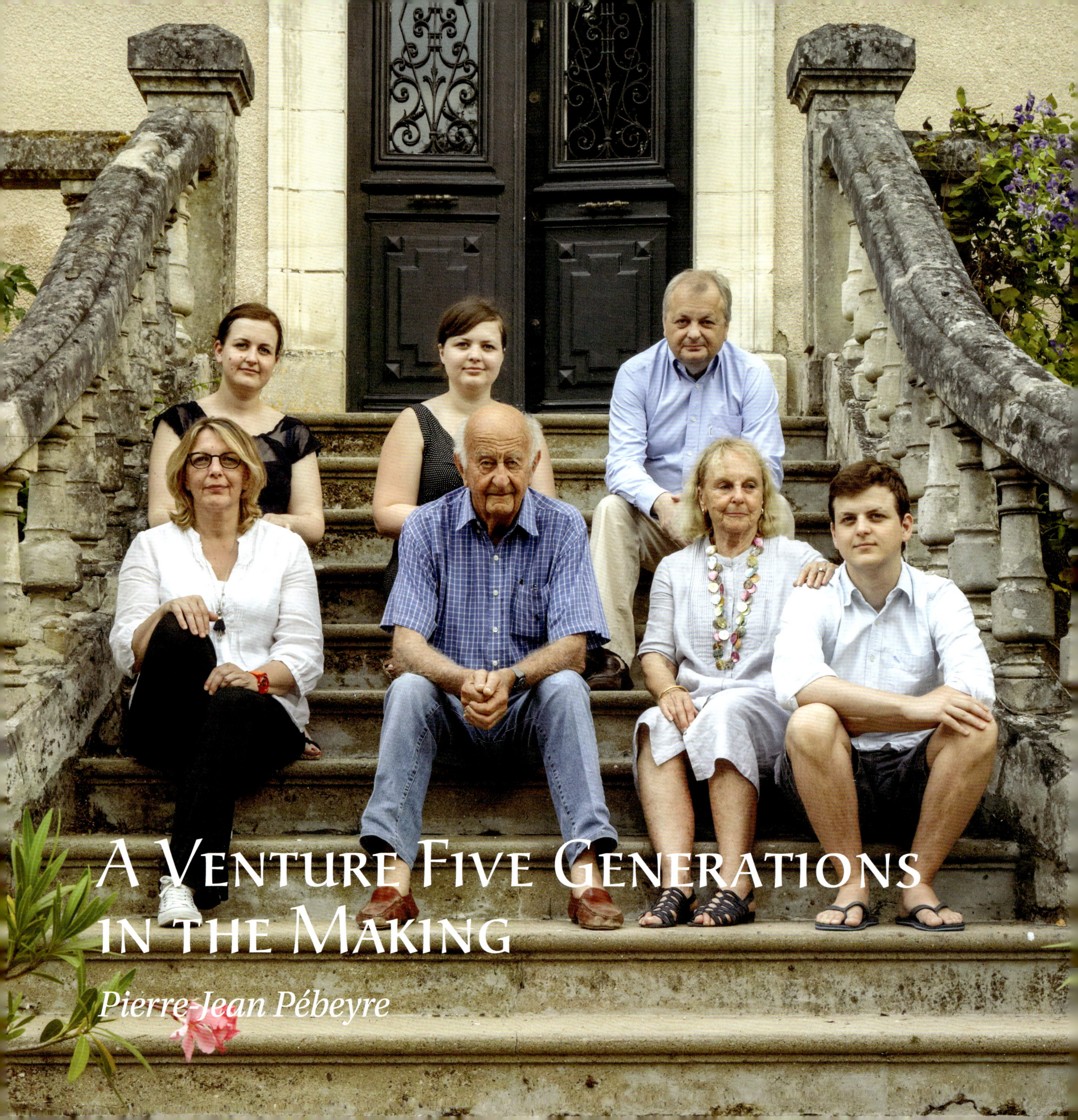

A Venture Five Generations in the Making

Pierre-Jean Pébeyre

The birth of the Pierre Pébeyre business

It was in 1897 that my great-grandfather, Pierre Pébeyre, founded his business. Before then he had been a schoolmaster, although truffles were his calling.

ONE MAN'S AUDACITY

In class, his mind would drift to the truffle orchards. Before long, he abandoned school to devote himself to his passion: the truffle business. After having learned his new trade in the region's specialist houses, he decided to set up on his own, in his small village of La Chapelle-Mareuil in the north of the Lot. For years he filled his books of accounts and letters with his beautiful schoolmaster's handwriting.

At the beginning, in addition to fresh truffles, Pierre Pébeyre sold all sorts of "little products", local specialities produced in the surrounding countryside: for example, ceps, walnuts, asparagus, etc. Foie gras made a significant contribution to sales, especially foie gras with truffles. My great-grandmother followed the nineteenth-century trend and created her own sideline of conserves, encouraged by the expansion of the canning process, which provided more regular sales. The truffles were brushed by hand, sorted and then sterilized in thick-glassed conical flasks and sealed with corks and metal wire. The new potential for conservation and delivery made it possible, little by little, to select only the best truffles

When Pierre Pébeyre began trading, truffle production had already peaked. An abundance of truffles was available and while truffles were an exceptional food, they were not yet a luxury one, and the wealthy were not the only people able to consume them. France produced an average of 800 tonnes of truffles per year. Production began to decline in Perigord, but remained high in Quercy and the southeast of France. In the Lot, only the Mareuil region produced a good hundred tonnes a year.

The tubers gathered at this time most often came from wild truffle orchards. Uncultivated land was not left idle; woodcutters and shepherds passing through the woods and thickets created good conditions for wild truffle growth. Firewood cutting cleared the vegetation cover; sheep weeded the soil and enriched the organic matter. Looking after truffles was part of the shepherds' work. Finally, in the still densely populated countryside, berry picking and mushroom hunting were widespread because they were profitable. You never sat on your hands or stopped thinking about things. Sons, daughters, grandmothers, all collected dead wood, poked about under trees and scratched around in the soil ... all while looking after cows, sheep or geese. Everyone brought something back, be it chestnuts, medlars or hazelnuts in their pockets or ceps and truffles in their aprons. People were hunter-gatherers as much as farmers. Everyone had a method for detecting truffles and was guided by certain indicators, such as a fly's flight or a circle of parched grass around a tree.

Previous pages: The village de Saint Cirq Lapopie, in the truffle-producing region of the Lot.

Facing page:
Front row from left to right: Babé, Jacques, Monique and Pierre Pébeyre.
Back row from left to right: Marie, Cecile and Pierre-Jean Pébeyre.

Located at the heart of one of the biggest regions of production, my great-grandfather bought from local markets in Quercy and Perigord's small towns such as Martel, Gourdon and Sarlat. He also had truffles brought in from the southeast of France which he bought, especially, from Monsieur Martin, a broker. At his request, the latter sent him a young nephew who was charged with doing the small markets. This was Jean-Marie Valayer, who later created the largest French truffle brokerage house, still run by his descendants today.

Sometimes it is not easy to stop a pig from eating a truffle he has found! Late-nineteenth-century print.

A late-nineteenth-century postcard showing truffles being prepared in a factory in Périgueux.

A time of plenty

The great proliferation of truffles in France was primarily due to two natural disasters: the phylloxera that ravaged vineyards between 1868 and 1872, and floods.

Before the parasite's attack, winegrowers regularly fought to prevent wild truffles from invading their vines. However, after the phylloxera epidemic the vines had to be pulled up. Under duress, ex-winegrowers planted oaks as a replacement for the missing vines, thus presenting truffles with optimum growing conditions. As for the south-east of France, a substantial campaign of reforestation was embarked upon during the Second Empire in Barronnies, the Basses-Alpes and the Mont Ventoux regions following disastrous floods. Here too, thousands of oaks were planted, this time to prevent the soil from being washed away by heavy flooding.

The results of these plantations for the proliferation of truffles were comparable to what was being produced in Perigord and Quercy at that time. Added to the existing truffle orchards, the new trees started to produce little by little, the last trees beginning to yield truffles at the beginning of the twentieth century.

At this time the rural population was still great and there was no problem with labour, and the tradition was still thriving.

TRUFFES EN GROS
DES MEILLEURES PROVENANCES
CONSERVATION GARANTIE
TRUFFES BROSSÉES EXTRA
PRODUCT OF FRANCE
Ingredients: Truffles, salt, truffles juice.
Net contents: 7 OZ
Ingrédients : Truffes, jus de truffes, sel.
425 ml
Poids net
230 g
Poids net égoutté
200 g
A consommer de préférence avant la date figurant sur le couvercle
P. PÉBEYRE
CAHORS (LOT)

TRUFFES EN GROS
DES MEILLEURES PROVENANCES
CONSERVATION GARANTIE
TRUFFES BROSSÉES 1er CHOIX
PRODUCT OF FRANCE
Ingredients
Truffles, Salt, Water
NET CONTENTS : 7 OZ
Composants
Truffes, Jus de Truffes
Sel
POIDS NET : 200 g
P. PÉBEYRE
CAHORS (LOT)

Alain Pébeyre and moving to Cahors

My great-grandfather Pierre, while he was passionate about the trade, harboured no great ambitions for his business. In contrast, he had great ambitions for his son, Alain. He decided, unusually for the time, to send the young man to study business in Switzerland.

In the aftermath of the First World War, Alain returned from his captivity to become head of the Maison Pébeyre. The studies his father had made him do determined him to expand the company.

The war had been the first break in truffle production. The truffle-producing trees planted at the end of the preceding century had to be replaced, but the renewal was only partially done. In the countryside that had been bled dry, where the urgent need was to feed the population, farmers concentrated on food crops to the detriment of less essential production. Although the volumes harvested remained high, truffle production had already begun to decline. The value of truffles gradually increased and paradoxically interest in the product continued to grow.

Confident in this evolution, Alain chose to devote himself to selling truffles and to abandon foie gras and ceps. Henceforth, the name Pébeyre would exclusively be associated with truffles. This decision differed from his father's commercial policy and also those of his competitors, who instead persisted on expanding their ranges. But by keeping his finger in just one pie, Alain ensured the reputation of the Maison Pébeyre, which became a benchmark in its domain.

Labels for tins of truffles from the Maison Pébeyre. They have not changed since 1897.

In 1920, the business was relocated from its original hamlet and set up in the centre of Cahors, the main town of the department of Lot. There it benefited from the town's ideal infrastructure. Modern means of communication, beginning with the telephone, were essential to receive and make orders. Close by, the post office and station provided the means for fast and secure truffle delivery.

While declining in the Dordogne and the north of the Lot, truffle production remained strong in the Cahors region. However, the need for the business to be based at the heart of the region of production was no longer the imperative it was thirty years earlier. As many truffles were supplied from the south-east of France as from the south-west of France. The company had an organized network of brokers, and buyers from the house kept abreast of each market. Truffles were also imported from Italy.

The transatlantic era

Long reserved for the great tables of Europe, the truffle became democratized. Its clientele expanded to companies from the charcuterie industry. The many small local charcuteries spread the taste of truffles to a wide range of consumers. Alsace foie gras houses used large quantities of truffles; a foie gras d'oie was inconceivable without them. Most of the time, the tubers were peeled, because truffles in their skins would mean the foie gras could not be served with a spoon, which was customary at the time.

Booming maritime links provided passengers with an opportunity, in order to prevent the boredom of such long liner journeys, to discover French haute cuisine, in

Label for a jar of truffles from the 1950s destined for the United States.

which the truffle reigned supreme. Having arrived at their destinations, travellers sought out what they had enjoyed during their crossings. By promoting the truffle, the transatlantic companies played a major role in the development of its consumption in the United States.

In Cahors, Alain Pébeyre profited from this development and arranged a strong line of business with North America. Other destinations, such as Spain, seem more surprising. How could the business enjoy a large market in Madrid in the 1920s? It was thanks to a great Spanish broker in love with the aroma of the truffle.

At that time, time was not money. On each of his business trips to Paris, Alain would not think twice about setting aside evenings for the theatre or the opera. There was no production line of professional meetings, and train journeys across Europe took several days.

In the factory in Cahors, work had changed a great deal. While the sorting of fresh truffles remained unchanged, their preparation and packaging had been mechanized. Brushing by hand, which caused big problems for the organization, was abandoned in favour a machine specially designed for cleaning truffles. Glass bottles and cooking in a bain marie were no longer used. They were replaced by stainless steel cans whose tighter seals allowed cooking by immersion at higher temperatures. Sterilization of the cans was done in autoclaves, while the large storage containers were still sealed with tin solder.

The growing demand for peeled truffles required more work, which was done in the factory and also at home. The "peelers"—the operation was entrusted to women—took home a certain quantity of truffles and had to bring back a roughly equivalent weight of peeled truffles and peelings. Needless to say, this "roughly equivalent weight" was the subject of long animated discussions, in which, it seems, my grandfather proved to be formidably severe.

A CENTURY OF UPHEAVALS

Truffles become scarce

Since the First World War, there has been a rural exodus from the French provinces. Country folk left to find work in the towns. Entire lands have been deserted, abandoned. With no one left to maintain the undergrowth, truffles could no longer form. In addition, the fewer peasants who remained were more attentive than ever to the profitability of their investments. As a result, truffles were relegated to the poorest land in favour of more immediately profitable crops. In the depopulated villages, the once common small truffle markets have gradually disappeared.

Since the 1940s, the decline in production has accelerated. Farms were mechanized, with dramatic consequences for the fungus because the machines destroyed the superficial roots of their host plants. Truffle oaks that produced truffles in their heyday were not replaced—the productive life of a truffle oak is no more than sixty years. In addition, around Cahors, former hillside vineyards where the vines had been replaced by oaks suffered from soil degradation, as the soil was no longer enriched by *colombine*, a fertilizer formerly provided by pigeons. During this time the rural exodus continued inexorably.

A truffle cleaning machine purchased by Alain Pébeyre in 1920.

A new clientele for Jacques Pébeyre

All these problems have intensified since the Second World War, when my father, Jacques Pébeyre, took over the family business. In rural areas there has been a radical change in attitudes and culture. Peasants have become farmers. Farming mechanization became widespread and farm management modernized. All these efforts focussed on crops that could ensure France's independence. "Small products", like the truffle, became increasingly marginalized. In the 1950s, wild truffle trees declined, the old truffle orchards ceasing producing while more recent plantations no longer received the necessary attention. Truffle production collapsed for the first time, falling from 300 to 400 tonnes annually to fewer than 150 tonnes.

Truffles became prohibitively expensive. A substantial number of food shops stopped selling them. The low volume of supply and high price prevented the incorporation of truffles in dishes for the general public. However, the truffle's scarcity reinforced its reputation. Jacques Pébeyre saw an opportunity to develop a new clientele of great restaurateurs in France and abroad (England, the United States, Japan and elsewhere). This policy would gradually ensure the principal outlet for the company and secure it a major role in the market.

However, faced with falling French production, Jacques went in search of new sources of supply. He intensified commercial relationships with other countries, such as Italy and above all Spain. The presence of truffles in the Iberian peninsular had been known about since the end of the nineteenth century. Some French brokers had attempted to establish commercial links without great success. But at the beginning of the 1960s, Spanish traders discovered this new fungus and understood what profits could be made from it. Jacques thus undertook to travel through Spain and help producers organize the harvesting and sale of their truffles. Regions of production were discovered one after another from upper Aragon to Guadalajara. With the opening of this new market, the consistent supply of quality truffles enabled the business to compensate for the drop in French production.

Early experiments with cultivation

In the two old truffle-producing countries, France and Italy, the truffle's gradual disappearance has become a preoccupation. The volumes harvested have continued to decline. Sadly, as a result of the rural exodus, a large amount of truffle knowledge has been lost. How might this knowledge be recovered and decades of experience be regained?

In 1965, the first scientific research on truffles was done by French and Italian teams. In the mid-1970s, their work led to the distribution of mycorrhizal trees, which could provide truffles. Keen to test these trees and monitor their effectiveness, Jacques, accompanied by his friend Jean Rougié, got together with a nurseryman to develop a twelve-hectare experimental truffle orchard. At the same time, he initiated an ambitious research programme on the physiology of the truffle, with the help of another friend, the mycologist Charles Montant, a professor at the University of Toulouse.

Ever since, researchers have been working tirelessly to resolve the problems. Studies of the natural environment have regularly provided information on the conditions for the fungus' growth and possible opportunities for it cultivation. This progress in knowledge has nourished hope that one day we might count on increased production, which will bring truffles within reach of all.

The Maison Pébeyre now sells a wide variety of truffle-based products, ranging from oils and vinegars to salt, pasta and, of course, foie gras.

STRINGOZZI
AUX TRUFFES
TAGLIATELLES
AUX TRUFFES
TRUFFES BROSSÉES
BLOC OF DUCK FOIE GRAS
SAUCE AU FOIE GRAS
TRUFFÉE
FOIE GRAS
AUX TRUFFES 10%
PEBEYRE
TRUFFES
Truffes
Truffes
SAUCE BLANCHE
AUX TRUFFES
SAUCE AUX TRUFFES
7,5%
100 ml

Pierre-Jean Pébeyre in an era of research

In 1983 it was my turn to join the business. Not long after this I met Ken Hom and he entered our family life.

At this point in our story, French truffle production had declined to a mere 35 tonnes per year. The precious fungus had become an exclusive product, reserved for a small circle of customers. The issue of truffle cultivation had still not been resolved. However, the research programme launched by the business did reveal plenty of secrets, both biological and chemical. Despite difficulties observing the plant, its life cycle was studied and described. Its unique aroma was decrypted, producing material for a patent application. A specialized fertilizer, also patented, was created and supplied to producers. The truffle was also the subject of six doctoral theses and many scientific publications. All this work, initiated and financed by the business, significantly advanced truffle knowledge.

Yet, despite the considerable contribution of research, an essential problem remained: that of the birth of small truffles. French truffle orchards failed to arrest the decline in production and yielded just 15 to 20 tonnes annually at the turn of the 2010s.

Against this background, the company has continued to adapt, expanding the business again. After the decades of hyperspecialization by the two previous generations, I revived my great-grandfather's trade and rediscovered the great tradition of foie gras with truffles. With the addition of other truffle products and ceps, the range was expanded to new products resulting from thirty years of research: truffle butter, truffle sauce, oil infused with truffles, etc.

Gift packaging for fresh black truffles.

With the same passion as my forefathers, I developed the international export market for truffles alongside this diversification, and became interested in new varieties, such as the Chinese truffle, as well as new areas of production that appeared in the southern hemisphere. With the development of the Internet, the company benefitted from a sales network that provided a new clientele looking for fresh truffles.

The business has resolutely remained a family business, positioned itself in new global markets, and continued to expand its product range in order to provide the fifth generation, Marie, Cécile and Pierre, with the means of continuing this wonderful venture.

Pierre-Jean (left) and Jacques Pébeyre (right) sorting fresh black truffles.

THE TRUFFLE IN ALL ITS GUISES

Pierre-Jean Pébeyre

Pigs and dogs

There are some people who are able to locate truffles underground simply by observing the quality of the ground and changes on its surface. For example, certain types of fly, attracted to the smell of truffles, lay their eggs just above or take flight when a hunter arrives. Scraping the soil away in this place, or under the circles the flies fly in, may uncover a truffle. But this method, which takes time and yields more unpredictable results, does not suit everyone.

The preferred method for hunting and harvesting truffles has been to use pigs or dogs. In the past, pigs were often used, simply because most peasants had at least one of them. Animals were killed in autumn every year and their replacements were therefore young in winter, which made the work easier, because a one- or two-year-old pig is not easy to handle and tires easily. Dogs, especially sheep-dogs, have also been widely used. If one opts for a pig, they have to be chosen young. The pigs are kept on a leash and pinpoint tubers with their snouts, but they have to be forcefully pulled back when they find a truffle because they will not hesitate to gobble it up. It is wise to reward them with some potato or corn each time they succeed, and there are also some pigs that have to have their backs scratched to be encouraged to work willingly.

Hunting with pigs is becoming increasingly less common, something I personally regret. They are

Left: Hunting for black truffles with a pig in winter.

Right: Print from 1893 showing a woman from Arcambal in the Lot with her truffle-hunting pig.

Truffle hunting with a dog in a truffle orchard near Cahors, South-West France.

intelligent animals blessed with a fine sense of smell. The truffle hunter and his or her pig is a cherished image from our past; they make a well-matched couple, frequently immortalized on postcards. A dozen years ago, in the Mercantour massif, a young boar was tamed and trained to hunt truffles. It is not so surprising that a wild pig might be as successful at this or even be better than domestic pigs. Wild boars have always adored truffles, and have been a real threat to truffle cultivators. They lay waste to truffle orchards, digging them up and destroying them. Traditional methods are used to keep them at bay, such as spreading mothballs to keep the boars away like big moths, or spreading sachets of hair—because they do not like the smell of man—but the rising population of boars as a result of hunting limitations has not helped solve this problem.

Dogs are increasingly preferred to pigs these days and their senses of smell are trained especially to hunt truffles, just as police dogs are trained to hunt drugs. A truffle-hunting dog will make a hole in the ground as soon as it finds a truffle, and naturally it too is given a reward for its find.

Although there is no breed of dog that is especially more talented at truffle hunting than any other, hunting dogs are known to be the least adept because they prefer to follow the scent of game rather than concentrate on truffles. Sheepdogs or plain mongrels are excellent collaborators.

Hunting with a dog for truffles in Spain. The bag contains harvested truffles and rewards for the sniffer dog.

CULTIVATING TRUFFLES

The cultivation of truffles in the same way that other fungi can be cultivated is still a dream, but the desire to try one's luck and plant one's own truffle-producing trees is, nevertheless, a very strong one. We do not yet know enough about the truffle, but we know sufficient to avoid committing some of the possible mistakes when planting.

1. **Choice of soil is important;** all truffle orchards in the world have highly calcareous soils with a pH of around 8. The soil should be rich enough in organic matter to be able to nourish a young truffle when it is just forming, because they are autonomous from birth and must therefore fend for themselves to find what they need to feed on and drink. In France, there is a fertilizer called Fructitruf that can help where the soil is too poor. While these soil requirements strongly limit scope for planting, climate needs present a further limitation.

2. **Climate choice:** Truffles can only grow properly in a temperate, Mediterranean-type climate. They need water and heat in spring, heat punctuated by large storms and heavy rain in summer, wet autumns free from heavy frosts and cold winters with frosts of no lower than –5 °C. A lack of water is very easy to resolve with irrigation. Heavy frosts, in contrast, are much harder to manage. This combination of climate and very specific soil requirements explains why our truffle can only currently be found in the north Mediterranean, in France, Spain and Italy. It may be possible to find these conditions in the southern hemisphere and perhaps in North America.

3. **Tree choice:** The choice of climate and soil alone are not enough to produce truffles; the tree also has an important role.

There are several species of tree that can encourage the birth and growth of truffles. The best known are oak, hazel, lime, the Aleppo pine and the hornbeam. But, of all these species, the one that is most commonly encountered is clearly the oak. For a tree to encourage truffle growth it must be mycorrhizal, mycorrhizae being a meeting point between the fungus that will produce a truffle and a tree. Scientific research over the last thirty years has produced trees bearing the truffle mycorrhiza, *Tuber melanosporum*. This long-known technique was perfected at the beginning of the 1970s and enables the production of very good quality trees that are very well mycorrhized. Sadly, this does not mean that all these trees will produce truffles. Planting is usually done using traditional tree-planting techniques.

Hunting for truffles in a truffle orchard planted with oaks near Cahors, South-West France.

TRUFFLE MARKETS

Each truffle-producing region has one or more weekly markets. Their season coincides with the truffle harvest: from the end of November to the end of March. These very traditional markets have existed since the nineteenth century. They are the only remaining agricultural markets in Europe where the forces of supply and demand operate outside any regulatory mechanism. The large fluctuations in prices from one week to the next have had consequences and the number of market has steeply declined as truffles have become increasingly scarce.

A late-nineteenth-century postcard showing a truffle market in the Dordogne.

In France, there are only four or five places of any real significance as opposed to forty or so at the beginning of the twentieth century. How these markets are organized has barely changed since they began and is the same for each location; producers bring the truffles they have harvested two to three days previously and sell them to the trading houses' brokers.

A broker buys all of a producer's truffles. They are not itemized and there is not enough time look closely at the quality of the lots offered. Not being able to select the truffles, brokers select by producer and rely on groups of producers they have known for years who bring them their truffles first. These are what the broker calls his "clientele". When each party has agreed on a price, the broker gives the producer a slip of paper on which he writes the price per kilogramme he has offered for the truffles. This paper serves as a contract: at the end of the market the producer has his truffles weighed by the buyer, and thcy arc thcn bought at the price written on the slip of paper.

Contrary to popular belief, there is a lot of trust in these markets; the participants know each other well and have generally done so for a long time. The brokers working for our house have done so for thirty or forty years; some, from father to son, have done so for nearly 100 years.

Above: A buyer weighing truffles using a steelyard, Uzès market in the Gard.

Below: Truffle collectors presenting their truffles to buyers in Lalbenque market in the Lot.

Jacques Pébeyre

How do you recognise a good truffle? Quality is a decisive criterion for choosing a truffle, but how do you recognise it? The method is simple, although it cannot be repeated often enough: you have to

QUALITY CRITERIA

look inside it. What a truffle looks like, under its skin, is of the utmost importance. It is essential to peel back a thin strip of skin. A strong contrast between a truffle's black flesh and the whiteness of its veins is a guarantee of quality and freshness, since the veins shrink as a truffle matures.

Using a sharp knife, cut a piece of truffle. Sometimes this has to be done in several places. The truffle's flesh should be firm and black, crisscrossed in white veins. The thickness of the veins depends on truffle variety, while the veins' colour depends on season. Thus, the later a truffle is harvested, the darker its veins. Woody or rotten truffles should obviously be rejected. Woody truffles, that is those that are hard like wood, are the result of a summer that was too dry, while rotten truffles are generally the result of a prolonged winter and frosts that were too severe. They become waterlogged, soften and very quickly spoil. Sometimes, only the upper part of a truffle has been frozen; in this case it is easy to remove this bit and only retain the best part. Of the rejected bits, those that are not rotten are set aside to make truffle breakings (*brisure de truffe*). In the worst case, frozen or rotten truffles are "sown" in truffle-growing land, that's to say next to host trees, in order to encourage the growth of new truffles.

Some claim they can determine a truffle's quality based on its smell alone. Sadly, this is possible only to a limited extent, and the surest way to obtain quality truffles it to visit a well-respected truffle house. The best truffles are harvested in January and February. In principle, by then they have reached their optimal maturity and developed their full flavour. On the other hand, be careful when buying fresh truffles before the second fortnight of December. There is a risk that these will not be perfectly ripe. Also, make sure they are neither too firm nor too soft. A truffle's ideal consistency is similar to that of a firm rubber ball.

A truffle and truffle knife sitting on a page of a book published in 1832 by Joseph Roques, the great mycologist and truffle enthusiast.

STORING AND PRESERVING TRUFFLES

Preserving truffles

Truffles are a very delicate food. In fact, they mostly consist of water, and this rapidly evaporates during storage. It is, therefore, recommended that they not be stored for long. Ideally they should be placed on a bed of rice. Tip the rice into a screw-top jar or a hermetically sealed plastic box and place the truffles on it in such a way that they are not touching each other. Close the jar or box and keep the container in a refrigerator. The ideal storage temperature is between +4 °C and +6 °C. The rice absorbs the humidity and enables fresh truffles to be kept for a week. Check their condition every day. A thin white film on the surface of the truffles indicates they should be consumed immediately. But this film, a harmless mould, is easily brushed off under running water. Incidentally, if you choose a good quality rice to preserve the truffles with (for instance an aged basmati rice), it will be delicious when simply steamed and accompanied with a knob of butter.

Truffles never taste as good as when they are fresh, but if you have too many (for which you should have no reason to complain), you can freeze them or can them. In both cases their white veins will disappear. Never put them in oil or alcohol. Oil spoils them and alcohol distorts their taste.

To freeze truffles, place them in individual freezer bags, in screw-top jars or in small plastic boxes. Using this method means that, while the defrosted truffles will certainly lose a little of their consistency, they will retain their flavour. Ideally, cut the truffles into pieces while they are still half frozen and immediately incorporate them into your recipes. Canning not only preserves truffles the longest but is also the most satisfactory.

Place a few truffles in a jar without squeezing them in. Add two or three tablespoons of water and a little salt. Hermetically seal the jar and place it in boiling water, boiling it for at least three hours. The truffles will lose around 25 percent of their weight during this process: the truffle juice produced in this process can obviously be reused in your cooking. As well as finding whole truffles in cans and jars, you can buy pieces, truffle peelings (*pelures de truffes*, coarsely ground second-choice truffles) or the more finely ground truffle breakings. The main problem with using truffles from cans or jars lies in deciding how much to use. It is generally difficult to know how much to use in your dishes to achieve a sufficiently pronounced taste. My advice is to follow this golden rule: allow for 12.5 g of truffles (the smallest can commercially available) per person and you will be sure of success.

Cooked black truffles preserved in glass jars.

COOKING WITH TRUFFLES

Truffles can be chopped, crushed, or thickly or thinly sliced. There are small adjustable mandolins available for this purpose, but a good sharp knife is just as good. The chef Jean Delaveyne advises that they should be crushed rather than cut into small pieces in order to release their flavour. Fresh truffles are in themselves a foodstuff and are suitable for cutting into thinish slices so they can be munched in all their splendour, in a salad for example.

While preserving truffles changes their flavour a little, it does not spoil it. Some claim that preserved truffles are not as good as fresh ones. I do not agree. Their flavour is clearly different but more complex in my opinion, with subtleties of its own. I would even say that fresh truffles are more suitable for some recipes and in some preparations while preserved truffles are more suitable for others.

Cold egg terrine with truffles.

Fresh truffles have a stronger flavour but also one that is a little wilder than that of preserved truffles. It is a forthright and earthy flavour that goes well with raw vegetables and other simple flavours. Sliced fresh truffles are delicious on grilled filet steak, lamb noisettes, lambs lettuce or a good quality poultry breast. On toast they make the most luxurious, refined and nevertheless simplest of toasted sandwiches. Preferably, it is fresh truffles that should used in truffle omelettes. For scrambled eggs preserved truffles are preferable; their flavour is more sophisticated than that of fresh truffles and better harmonizes with the creamy consistency of the eggs. They should also be used to stud foie gras terrine. Generally, preserved truffles are so delicate and so spiritual that they have to be found a support, a foil.

When fresh, truffles handle competition better, but this tolerance cannot be pushed too far. Indeed, whether fresh or preserved, the truffle is paradoxical when it comes

Terrine of foie gras layered with truffles.

to flavour. Here is one of the strongest flavours in our flavour repertoire and yet it is as fragile as it is strong. It has a pronounced and delicate flavour that can be bullied when in competition with other strong flavours. Ingredients to accompany it should always be chosen with care; if an element is too assertive, the flavour balance will be compromised. In the south-west of France, truffles have always been used with goose or duck fat, or with lard.

Meats, foie gras, potatoes and fresh vegetables all go deliciously well with truffles, as does one particular strong flavour: garlic. Each time you use truffles in your cooking, particularly when fresh, take care to run your knife through a garlic clove first. The affinity between garlic and truffles is remarkable.

Finally, never place truffles inside poultry, as was popular in centuries past: it amplifies the odour from the aromatic strength of the fungus and the result is unpleasant. Perhaps this may not have bothered our ancestors, who loved this gamey flavour more than we do. The best way to add truffles to poultry is the equally old method called *"demi-deuil"* in French, where a good quantity of truffle slices are slipped under the skin of the poultry and cooked with it: the result is incomparable.

Truffle sausages.

Chicken stuffed with truffles in the demi-deuil *style, ready for poaching.*

Classic Recipes

This is one of the simplest and most delicious ways to enjoy black winter truffles. These addictive sandwiches are a wonderful appetizer and perfect with champagne. Make them the day before so the bread can absorb as much of the truffle aroma as possible. Good bread, in addition to the truffles, is the key to this recipe.

Truffle Sandwiches (Tastou)

Serves 4 to 6

8 thin slices of country bread or whole wheat bread

200 g / 7 oz unsalted butter

80 g / 2¾ oz fresh black winter truffle, thinly sliced

Fleur de sel sea salt or coarse salt and freshly ground black pepper

The day before, butter each slice of bread generously on one side. Then cover with truffle slices and salt and pepper. Close with another slice of bread with the buttered side facing the truffles. Continue the process until you have four large sandwiches. Place them together, tightly wrapped in cling film under a weight so that the flavours absorb into the bread, and refrigerate.

The next day, just before cooking, preheat the oven to 220 °C/gas mark 7 for 15 minutes. Place the four sandwiches on an oven tray and cook for about 10–12 minutes until the bread is lightly toasted.

Remove, cut into serving size and serve at once.

Here is a simple, refreshing salad enriched with the best winter truffles of the season. When using truffles, it is best to use a neutral oil to avoid overwhelming their delicate flavour.

Fresh Truffle Salad

Serves 4

2 bunches of fresh lambs lettuce, washed and spin dried

Balsamic vinegar

Salt, to taste

Freshly ground black pepper to taste

Sunflower oil

60 g / 2 oz fresh black winter truffles, thinly sliced

In a large salad bowl, mix the vinegar together with the salt and pepper. Whisk in the oil. Save a small portion of the vinaigrette and put aside. Gently toss the salad with the vinaigrette.

Moisten the truffle slices with the vinaigrette you had set aside. Place them on top of the mixed salad and serve at once.

This is an unusual but delicious way to enjoy fresh black winter truffles. Coulommiers is a raw milk creamy cheese. It is thicker and less acidic than Brie, for example. Mascarpone is an Italian cheese that is fatty and rich. Its mild flavour allows it to absorb the strong truffle aroma. Easy and simple, this is a delightful after-dinner treat. Nothing could be easier to assemble.

Coulommiers Filled with Truffles

1 round of ripe Coulommiers cheese, about 400 g / 14 oz

125 g / 4½ oz mascarpone cheese

80 g / 2¾ oz fresh black winter truffles, thinly sliced

Take the Coulommiers cheese and split it in half with a sharp knife. You should have two whole round slices.

Spread the mascarpone cheese on the two inside halves of the Coulommiers, carefully avoiding spilling onto the edges.

Layer the truffles on one half of the cheese, cover with the other half. Press well.

Wrap the cheese in cling film and refrigerate for at least 2 days.

Two or 3 hours before you are ready to serve, bring the cheese to room temperature and serve with toasted bread.

This extravagant dish dates back to the time when truffles were abundant and were eaten as winter vegetables. This was originally a peasant dish that is well worth making if you want to truly appreciate the taste of truffles. In former times, when there were no home ovens, the truffles were placed in a copper casserole and the hot ashes from the fireplace were heaped over to cook the truffles in their own juices. Today, an oven works quite well for this recipe. The size of truffle served per person, of course, depends how extravagant you feel.

Pierre-Jean's Whole Roasted Truffle (Truffes sous les cendres)

Per person:

1 whole fresh black winter truffle weighing about 30–60 g / 1–2 oz

2 strips of pork fat or bacon

Freshly ground black pepper to taste

Squares of aluminium foil

Wrap the pork fat or bacon around each truffle. Season them with black pepper and individually wrap in aluminium foil.

Preheat the oven for 20 minutes to 220 °C/gas mark 7.

Place the truffles on an oven tray and roast for about 10–15 minutes, depending on the size of the truffles.

Remove from the oven and place on a warm plate. Serve at once.

This earthy winter vegetable soup epitomises the essence of truffle flavour. It is easy to make and much of the work can be done ahead of time. You will see this is a match made in heaven.

Babé Pébeyre's Celeriac Truffle Soup

Serves 4 to 6

50 g / 1¾ oz unsalted butter
2 small onions, finely chopped
100 g / 3½ oz shallots, finely chopped
350 g / 12 oz fresh celeriac root, peeled and chopped
250 g / 9 oz potatoes, peeled
10 g / 2½ teaspoons sugar
1.2 litres / 2 pints homemade chicken stock
Salt and freshly ground black pepper to taste
80 g / 2¾ oz fresh black winter truffles, julienne
3 tablespoons fresh chives, finely chopped

Heat the butter in a large pot. Stir-fry the onions and shallots gently without browning for about 5 minutes.

Meanwhile, cut the celeriac and potatoes into bite-sized chunks. Add them to the large pot together with the sugar, chicken stock, salt and pepper. Mix well, bring the mixture to a simmer and cook for 20 minutes.

Remove from the heat and allow to cool to a point where it can be handled.

Ladle the soup into a blender in small batches and process until completely smooth and creamy.

Slowly reheat the soup and stir in the truffles. Ladle the mixture into a soup tureen or individual bowls, garnish with fresh chives and serve at once.

One of the ideal ways to enjoy the full perfume of black winter truffles is in a clear chicken soup. The key to the success of this dish is a very good stock. The fresh truffles are shaved onto the hot broth. This method of limited cooking preserves all of the truffles' aroma and taste.

Clear Chicken Soup with Fresh Truffles

Serves 4

1.2 litres / 2 pints homemade chicken stock

Salt and freshly ground black pepper to taste

2 whole spring onions, finely shredded

80 g / 2¾ oz fresh black winter truffles

In a large saucepan, bring the stock to a simmer and add the salt and pepper. Cover the pot and remove from the heat. Allow to sit for 10 minutes.

Finally, stir in the spring onions.

Brush the truffles with care and thinly slice.

Ladle the soup into a large soup tureen or individual bowls, add the truffle slices and serve immediately.

This risotto dish that is slowly cooked to a smooth, creamy consistency is made by Babé Pébeyre a lot. She adds chopped fresh or tinned truffles and crème fraîche, which stops the cooking. The cooked risotto will nicely absorb the flavours and highlight the intensity of the aromatic truffle. Babé does not serve Parmesan cheese with the risotto as she feels it would clash with the truffle taste and flavours. After tasting this risotto, we think you will agree!

Babé Pébeyre's Truffle Risotto

Serves 4 to 6

- 75 g / 2¾ oz unsalted butter
- 50 g / 1¾ oz shallots, finely chopped
- 100 ml / 3½ fl oz dry white wine
- 250 g / 9 oz Italian Arborio rice
- 1.5 litres / 2½ pints homemade chicken stock
- Salt and freshly ground black pepper to taste
- 100 g / 3½ oz fresh winter black truffles or high-quality tinned winter truffle (cooked once), coarsely chopped
- 50 g crème fraîche

Heat a casserole dish until it is hot and add the butter. Then add the shallots and cook over a medium heat for 2 minutes until they are translucent but not browned.

Now add the white wine and continue to cook for 2 minutes or until most of the alcohol has cooked off. Finally add the rice and continue to stir for 2 minutes.

Turn the heat to low, add some of the stock and continue to stir until the rice has absorbed some of the stock. Continue to add stock and stir, allowing the stock to slowly evaporate until you have used up most of it. This will take about 25–30 minutes. Add salt and pepper to taste.

When the rice is cooked, stir in the chopped truffles and finally stir in the crème fraîche, mixing well. Serve immediately on a warm platter.

Truffles offer a source of inspiration for everyone's favourite food—pasta. The earthy flavours of the truffle are the perfect foil for the simple pasta. Here is a delectable pasta dish that is quick to cook and even more delicious to eat.

Pasta with Fresh Truffles

Serves 4

250 g / 9 oz of your favourite dried Italian pasta or fresh pasta, such as tagliatelle, fusilli or farfalle

50 g / 1¾ oz fresh winter black truffles or highest quality tinned winter truffle (cooked once)

75 g / 2¾ oz unsalted butter

Salt and freshly ground black pepper to taste

3–4 tablespoons grated fresh Parmesan cheese

Garnish:

2 tablespoons fresh chives or spring onions (green part only), finely chopped

Cook the pasta in a large pot of salted water, according to the instructions on the packet or your favourite method.

Meanwhile, thinly slice the truffles. Divide the truffles in half and finely chop half of them, reserving the other slices for garnish. If you are using tinned truffles, combine any juice with the butter as well. Set this aside.

When the pasta is cooked, drain it well and reserve the water. Immediately combine the pasta with the butter, chopped truffles, cheese and about 3 tablespoons of pasta water. Mix well and adjust seasoning with salt and pepper.

Garnish with the remaining truffle strips and chives or spring onions and serve at once.

This incredibly simple dish comes from Monique Pébeyre. It is very easy to make and is even better to eat. We have found that cooked tin truffles of the finest quality (cooked once) work best in this recipe. This dish makes a stunning first course for any meal.

Monique Pébeyre's Soft Boiled Eggs with Truffle Sauce

Serves 6

Drain the truffles and put the juice in a small pan. Then purée the truffles in a blender or food processor until they are a smooth paste.

Cut the butter into small dice. Now heat the truffle juice over a low heat, slowly adding the butter bit by bit, whisking all the while until the butter has been fully incorporated. Then stir in the truffle purée. Continue to heat but never allow the mixture to boil. Season to taste with salt and pepper. Set aside and keep warm.

In a large pot, bring water to a boil and cook the eggs for precisely 4 minutes.

Remove the eggs, drain carefully and serve with egg holders. Each diner opens his or her egg and pours a small amount of truffle sauce in the egg. This is eaten with strips of toast as well.

6 slices of toasted bread

50 g / 1¾ oz tin of the best quality winter truffles (cooked once)

70 g / 2½ oz unsalted butter

Salt and fresh ground black pepper, to taste

6 fresh farm eggs

This is a classic egg dish with truffles. The subtle taste and texture of the eggs allow the aromatic truffle to shine. It is a rich dish but fairly easy to make. All that is required is a bit of patience and truffles.

Scrambled Eggs with Truffles

Serves 4 to 6

- 100 g / 3½ oz fresh black winter truffles or highest quality tinned black truffles (cooked once)
- 12 eggs
- Salt and freshly ground black pepper to taste
- 60 g / 2 oz unsalted butter
- 25 ml / 1 fl oz liquid cream

Crush and break up the truffles with a fork and add to the eggs. Add salt and pepper and allow the mixture to sit for at least 1 hour at room temperature.

Make a bain-marie by first bringing a pot of water to a boil. Then heat 50 g / 1¾ oz butter in another pan small enough to fit over the pot of boiling water. Now add the egg and truffle mixture and continue to whisk over the boiling water until the mixture is creamy.

When the mixture has achieved the desired consistency, stir in the remaining butter and cream. Serve at once.

Monique Pébeyre has been making this foolproof omelette for over 60 years. Her formula is 2 eggs and at least 20 grams of fresh winter truffles per person. Truffles and eggs are a perfect match. The subtle taste of eggs enhances the aromatic perfume of the truffle. Heating to no more than 30 °C will bring out its optimum flavour.

Monique Pébeyre's Truffle Omelette

Serves 4

8 eggs
80 g / 2¾ oz fresh winter truffles, sliced
Salt and freshly ground black pepper to taste
3 tablespoons cold water
60 g / 2 oz unsalted butter

Crack the eggs into a large bowl and add the truffles. Allow the mixture to sit for 1 hour or more at room temperature.

When you are ready to make the omelette, beat the eggs lightly and add salt and pepper. It is important not to beat the eggs too much.

Melt the butter in a nonstick pan. When the butter is hot, pour the egg-truffle mixture into the centre of the pan and cook over a medium heat, allowing the eggs to set and curl at the edges. Then, with a spatula, stir the eggs so the running part fills the areas towards the centre.

When most of the eggs are set and still slightly runny inside, fold one-third of the side of the omelette to the centre. Invert onto a warm platter and serve immediately.

This is a Bordeaux-inspired meat-wrapped treat. It uses crépinette, which is the French term for a small flat sausage, generally made of minced meat with chopped parsley and wrapped in caul (crépine), which can be obtained from your butcher. The lacy wrapping keeps the meat moist and juicy. These croquettes are wonderful with a light green salad.

Bordeaux Style Pork Croquettes with Truffles

Serves 4

- 225 g / 8 oz caul fat* (*crépine*)
- 450 g / 1 lb minced fatty pork
- 2 egg whites
- Salt and freshly ground black pepper to taste
- 2 teaspoons cognac
- 15 g / 1 oz fresh chives, finely chopped
- 5 g / ¼ oz parsley, finely chopped
- 13 g / ¾ oz garlic, finely chopped
- 100 g / 3½ oz fresh black winter truffles or high-quality tinned black winter truffles (cooked once), coarsely chopped
- flour, for dredging
- 2 eggs, beaten
- 35 g / 1¼ oz dried bread crumbs
- peanut or groundnut oil, for frying

Soak the caul fat in a bowl of cold water for 20 minutes. This will allow the caul fat to unravel easily.

In a large bowl, combine the pork with the egg whites, salt, pepper, garlic, cognac, all the herbs and the chopped truffles, and mix well.

Remove the caul fat from the water and pat dry with a tea towel. Cut the caul fat into four equal pieces. Lay out a square of caul fat and place a quarter of the pork mixture in the centre. Fold the sides and each end to form a package. Repeat until you have used up the entire mixture.

The crépinettes can be made up to 3 hours ahead of time, refrigerated and wrapped with cling film until ready to cook.

When you are ready to cook the crépinettes, dust them with flour, shaking off any excess, then dip in the beaten egg and finally dust them with the bread crumbs.

Heat a frying pan until it is hot, add the oil and, at the lowest possible heat, slowly pan-fry the croquettes on each side for 8 minutes until golden brown and crispy. Serve at once.

This is our version of a classic Escoffier recipe—chicken in half mourning (poulet demi-deuil).

We use a large organic chicken which we poach and steep in a classic Cantonese style. Steeping is an ideal technique for delicate foods, such as chicken and truffles. Here the gentlest possible heat is used so that the flesh of the chicken remains extremely moist and flavourful with a satiny almost velvet-like texture. The chicken simmers in liquid for a few minutes, then the heat is turned off, the pot tightly covered and the chicken left to steep to finish cooking. Adding ginger and spring onions gives the dish a Chinese touch without straying too far from its classical French origins.

Poached Chicken with Truffles (Poulet Demi-Deuil)

Serves 4–6

1 × 1.5–1.75 kg (3½–4 lb) organic farm chicken

Salt and freshly ground black pepper, to taste

6 slices fresh ginger

6 whole spring onions

50 g / 1¾ oz fresh black winter truffles or tinned high-quality winter truffles (cooked once), thinly sliced

For the sauce:

150 ml / 5 fl oz homemade chicken stock

20 g / ¾ oz unsalted butter, cut into bits

50 g / 1¾ oz fresh black winter truffles or tinned high-quality winter truffles (cooked once), cut into matchsticks

Salt and freshly ground pepper to taste

Sprinkle the salt and pepper evenly throughout the interior cavity of the chicken. With your fingers, loosen the skin from the breast. Gently stuff the thinly sliced truffles between the skin and the breast. Close the body cavity with a bamboo skewer.

Fill a pot, big enough to fit the entire chicken, with water and bring it to a boil. Add the chicken, salt, pepper, ginger and spring onions, tipping out any excess water. Bring the mixture to a simmer, cover and cook for 20 minutes. Turn off the heat and leave the cover on tightly and let it sit off the heat for 1 hour. After an hour, remove the chicken to a chopping board to cool.

Now make the sauce.

Strain and boil down the remaining liquid until you have 150 ml / 5 fl oz left. Slowly incorporate the butter into the liquid, then add the truffle matchsticks and salt and pepper to taste.

Carefully carve the chicken and put the pieces into the sauce to reheat slowly. Serve at once.

This recipe is inspired by Monique Delsol, whose nickname is Momo. She is one of the best cooks we know in our region of the Lot, in South-West France. Like many people from the area, she is a great lover of truffles and uses them whenever she can. Here Momo makes an incredibly savoury stuffing for inside the bird and under the skin. The result is one of the most delicious roast birds you will ever taste in your life.

Momo's Roast Guinea Fowl with Truffle Stuffing

Serves 4–6

1.5–1.7 kg / 3½–3¾ lbs guinea fowl

Salt and freshly ground black pepper

Truffle Stuffing:

100–110 g / 3½–3¾ oz fresh duck foie gras cut into large cubes

25 g / 1 oz finely chopped shallots

75 g / 2¾ oz minced sausage meat

75 g / 2¾ oz dried stale bread, broken into large chunks

50 g / 1¾ oz fresh black winter truffles or high-quality tinned black winter truffle (cooked once)

Salt and freshly ground black pepper to taste

Combine the salt and pepper. Sprinkle this mixture evenly throughout the exterior and interior cavity of the guinea fowl. With your fingers, gently loosen the skin from the breasts and thighs.

Combine all the ingredients for the truffle stuffing, seasoning well with salt and pepper. Put a small amount of stuffing under the skin and the rest inside the body cavity of the guinea fowl. Close the body cavity with a bamboo skewer.

Preheat the oven to 240 °C/gas mark 9. Meanwhile, place the guinea fowl on a roasting rack in a roasting pan, breast side down. Now put the guinea fowl in the oven and roast for 10 minutes. Then turn the heat down to 180 °C/gas mark 4 and roast for a further 30 minutes. Now turn the guinea fowl over breast side up and continue to roast for 10 minutes.

Remove the guinea fowl from the oven and let it sit for at least 15 minutes before carving. Carefully remove the skewer and take out the stuffing and set aside. Using a sharp knife, cut the guinea fowl into serving portions and arrange these on a warm serving platter together with the stuffing.

Chicken breasts have a mild subtle flavour that is a perfect foil for truffles. In fact, they are a wonderful duo. Here is a quick, easy and elegant way to make the most of your truffles.

Chicken Suprême with Truffles

Serves 4

4 boneless and skinless chicken breasts

80 g / 2¾ oz fresh black winter truffles or high-quality tinned black winter truffles (cooked once), sliced thickly

Salt and freshly ground black pepper to taste

60 g / 2 oz unsalted butter, cut into small cubes

Preheat the oven to 240 °C/gas mark 9.

Using a sharp knife, make deep incisions in each chicken breast. Insert slices of truffles throughout. Season generously with salt and pepper.

Place the chicken in an oven-proof pan. Add the butter on top of the chicken and cook for about 10 minutes. Remove from the oven and allow them to sit for at least 10 minutes before serving.

This is our version of a classic roast beef, served with a simple truffle sauce. Although fillet tends to be bland, it is a perfect foil for the delicate aromatic truffle. You will find this recipe ideal for an elegant meal, as it requires only a little work.

Roast Beef with Classic Truffle Sauce

Serves 4 to 6

1.3–1.5 kg / 2¾–3¼ lb beef fillet, trimmed and tied

2 tablespoons groundnut or vegetable oil

Salt and freshly ground black pepper to taste

Sauce:

3 tablespoons Madeira

50 ml / 1¾ fl oz demi-glace sauce (this can be made by reducing stock to a thick syrupy state or it can be purchased at fine gourmet shops)

60 g / 2 oz fresh black winter truffles or high-quality tinned black winter truffles (cooked once)

Preheat the oven to 220 °C/gas mark 7.

Season the fillet well with salt and pepper. Heat a roasting pan on top of the stove or hob. Then add the oil and slowly brown the fillet on all sides until it is a golden brown.

Put the beef in the oven for 10 minutes. Then reduce to the heat to 190 °C/gas mark 5 and cook for a further 10 minutes. Remove the beef. Let the roast rest at room temperature for 20 minutes before carving.

Chop the truffles for the sauce.

Pour off all the fat from the roasting pan, saving any bits and juices.

Then place the roasting pan back on the stove and deglaze the juices with the Madeira, and add the demi-glace and chopped truffles. Serve the fillet with the sauce.

The subtleness of veal is perfect with truffles. Its mild and delicate taste shows off truffles at their best. This is a sumptuous and elegant dish.

Roast Rack of Veal

Serves 4–6

Ask your butcher to prepare a rack of veal for roasting.

Preheat the oven to 220 °C/gas mark 7. Season the veal generously with salt and pepper.

Cut 50 g / 1¾ oz truffles into small thick pieces. Make deep cuts throughout the top of the rack of veal with a small sharp knife. Now insert the pieces of truffles in the cuts to stud it.

Put the veal in the oven for 10 minutes. Then reduce to the heat to 190 °C/gas mark 5 and cook for a further 25 minutes. Remove the veal and let it rest at room temperature for 20 minutes before carving.

Chop another 50 g / 1¾ oz truffles for the sauce.

Pour off all the fat from the roasting pan, saving any bits and juices.

Then place the roasting pan back on the stove and deglaze the juices with the Madeira, add the demi-glace and chopped truffles. Serve the veal with the sauce.

1–1.2 kg / 2¼–2¾ lb rack of veal

50 g / 1¾ oz fresh black winter truffles or high-quality tinned black winter truffles (cooked once)

Salt and freshly ground black pepper to taste

Sauce:

3 tablespoons Madeira

150 ml / 5 fl oz demi-glace sauce (this can be made by reducing stock to a thick syrupy state or it can be purchased at fine gourmet shops)

50 g / 1¾ oz fresh black winter truffles or high-quality tinned black winter truffles (cooked once), julienne

The humble potato is a natural foil for truffles. In this recipe, we present a delicious purée that is easy to make and will surely surprise your dinner guests and family. The cucumbers lighten the potatoes as well as giving it an unusual taste and texture that marries well with the earthy truffles.

Potato and Cucumber Purée with Truffles

Serves 4–6

Peel the cucumbers, slice them in half lengthways, and using a teaspoon, remove the seeds. Then cut the cucumber halves into slices. Sprinkle them with the salt and mix well. Put the mixture in a colander and let it sit for 45 minutes to drain. This rids the cucumber of any excess liquid. When the cucumber slices have drained, rinse them in water and purée in a blender or food processor. Then squeeze any excess moisture from the puréed cucumbers in a linen kitchen cloth. Set aside.

Peel and cook the potatoes in salted water for 20 minutes or until tender.

When the potatoes are cool enough, peel them, and pass them through a ricer or food mill. Reheat them in a large heavy pan. Bring the cream to a simmer in a sauce pan and whisk it into the potatoes. Incorporate the butter, mixing well. Add salt and pepper to taste and set aside.

Meanwhile, chop the truffles finely.

Just before serving, reheat and fold the cucumbers and truffles into the potato purée and serve at once.

1.5 kg / 3¼ lb cucumbers

2 tablespoons salt

900 g / 2 lb potatoes

250 ml / 9 fl oz liquid cream

125 g / 4½ oz unsalted butter

Salt and freshly ground black pepper to taste

80 g / 2¾ oz fresh black winter truffles or high-quality tinned black winter truffles (cooked once)

This is Jacques Pébeyre's version of a local delicacy usually made with cepes. Being a master of truffles, he uses fresh truffles, which marry extremely well with the earthy potatoes. It is a match made in heaven and unusually delicious too.

Jacques Pébeyre's Potato Sarladaise

Serves 4 to 6

1 kg / 2½ lb potatoes

2 tablespoons peanut or groundnut oil

Enough duck or goose fat (can be purchased in tins)

Salt and freshly ground black pepper to taste

60 g / 2 oz fresh black winter truffles, thinly sliced

30 g / 1 oz coarsely chopped garlic

30 g / 1 oz coarsely chopped fresh parsley

Heat a nonstick frying pan and add the oil, then the potatoes. Lower the heat to medium and continue to cook for 30 minutes. From time to time add a spoonful of duck or goose fat to keep the potatoes moist.

Lower the heat to very low, cover loosely and continue to cook for about 10 minutes, adding fat as needed. Uncover and continue to cook for about 30 minutes.

Invert the potatoes on a flat lid and gently slide the potato cake back into the pan so that it can brown on the other side. Add fat as needed.

When the potato cake is crispy and brown add salt, pepper, sliced truffles, garlic and parsley. Slide unto a warm platter and serve at once.

A crisp salad is the perfect complement to this dish, which is also a delicious accompaniment to roast meat

Fresh scallops are best cooked by the simple technique of steaming. Using hot wet vapours, this method brings out the subtle taste and succulent texture of scallops without overcooking them. This makes them a perfect companion with truffles. Here their natural juices are mixed with fish stock and finished with julienne truffles to make a tasty dish that is both simple and elegant. The small amount of butter adds a touch of richness and savouriness. It is ideal for a quick and easy dish and I think it makes an ideal opener for any dinner party.

Steamed Fresh Scallops with Truffles

Serves 4

450 g / 1 lb fresh scallops, including the corals

Salt and freshly ground black pepper to taste

50 ml / 1¾ fl oz homemade fish stock

2 tablespoons unsalted butter, cut into small pieces

80 g / 2¾ oz fresh black winter truffles or high-quality tinned black winter truffles (cooked once)

1 tablespoon finely chopped chives

Place the scallops evenly on a heat-proof plate. Sprinkle evenly with salt and pepper.

Next set up a steamer or put a rack into a wok or deep pan and fill it with 5 cm / 2 in of water. Bring the water to the boil over a high heat. Carefully lower the scallops with the plate into the steamer or onto the rack. Turn the heat to low and cover the wok or pan tightly. Steam gently for 5 minutes.

Meanwhile, finely julienne the truffles.

Remove the scallops with the plate from the steamer and pour off any liquid that resulted from the steaming into a saucepan. Add the stock and reduce by three-quarters. Slowly mix in the butter, return the scallops to the sauce to warm together with the julienne truffles, garnish with the chives and serve at once.

East–West Recipes

More than three decades ago I wrote extensively about a growing trend I then called East meets West cuisine, now popularly known as Pacific Rim or Fusion cooking. It is a cooking style that emphasizes the blending of foods, spices, flavourings and techniques formerly isolated from each other within Asian or European kitchens. This salmon dish is a typical example. Here I use rice paper (a very Asian ingredient) to wrap salmon (a very European fish), seasoning it with both fresh Asian and Western herbs and black truffles. It is surprisingly easy to make and the result is a delightful crackling and stunning-looking appetizer that will surely impress your family, friends and guests.

Crackling Rice Paper-Wrapped Salmon with Truffles

Serves 8

- Salt and freshly ground black pepper
- 900 g / 2 lb boneless, skinless salmon fillet
- 100 g / 3½ oz fresh black winter truffles or high-quality tinned black winter truffles (cooked once)
- 1 packet 22 cm / 8 in Bánh Tráng dried rice paper rounds
- 3 tablespoons chopped chives
- 3 tablespoons extra virgin olive oil

Combine the salt and pepper. Divide the salmon into 8 equal pieces, about 7.5 cm × 7.5 cm each. Then split each salmon piece in half. Sprinkle the salt and pepper combination on the inside and place slices of black truffles in between each salmon piece. Sprinkle the rest of the mixture evenly over the salmon pieces.

Fill a large bowl with hot water and dip one of the rice paper rounds in the water to soften. This will take but a few seconds. Remove and drain on a linen tea towel. Place a slice of truffle in the centre, then a piece of salmon filled with truffles. Then sprinkle some chopped chives over the salmon. Fold the first edge over the ingredients, then fold in the sides. Then fold the entire package in half, leaving a flap at the corner facing away from you. Finally push the flap to secure the package. Repeat until all the packages are filled.

Heat a large heavy frying pan over a high heat until it is hot, then add the olive oil. When the oil is hot, add the packages and pan-fry with the seamless sides down for about 3 minutes or until they are golden brown. Turn them over and continue to cook until golden brown.

Arrange them on a platter and serve at once. The dish is perfect with a small salad on the side.

This a savoury East-West starter that can inspire any meal. We find this delectable recipe works extremely well with high-quality tinned black winter truffles, and if you don't have time to make the Asian duck confit (see overleaf), you can easily substitute French duck confit. The rolls are so tasty, you will find that you do not even need a dipping sauce. They are perfect served with a simple green salad.

25 g / 1 oz bean thread (transparent) noodles

250 g / 9 oz Asian or French duck confit meat from thighs

225 g / 8 oz sausage or minced pork meat

1 small onion, finely chopped

50 g / 1¾ oz garlic, coarsely chopped

28 g / 1 oz shallots, finely chopped

6 g / ¼ oz chives, chopped

100 g / 3½ oz high-quality tinned black winter truffle (cooked once), coarsely chopped

Equal portions flour mixed with water

1 packet rice paper wrappers

400 ml / 14 fl oz vegetable oil, preferably groundnut

Salt and freshly ground black pepper to taste

Asian Duck Confit Spring Rolls with Truffles

Makes about 30 spring rolls

Soak the noodles in a large bowl of warm water for 15 minutes. When they are soft, drain them and discard the water. Cut them into 7.5 cm lengths using scissors or a knife.

Strip the meat from the thighs and coarsely chop until you have the 250 g / 9 oz required. Coarsely chop the duck skin and crisp the skin slowly in a wok or frying pan. Drain on kitchen paper and allow it to cool thoroughly. Pour off most of the duck fat, keeping a small amount in the wok or pan. Cook the pork with the remaining duck fat, adding the onions, garlic, spring onions and shallots. Cook the mixture for 10 minutes, then drain in a colander and allow it to cool.

When the pork is cool, combine it in a large bowl with the bean thread noodles, duck meat, crispy skin, salt and pepper, truffles, and chives. In a small bowl, mix a small amount of flour with water until you have a paste.

When you are ready to make the spring rolls, fill a large bowl with warm water. Dip one of the rice paper rounds in the water and let it soften. Remove and drain it on a linen towel. Put about two tablespoons of the filling on the bottom third of the softened rice paper wrapper. Bring edge nearest you up and over the filling, then tuck edge under filling. Fold in at the side and then roll up tightly. Seal the ends with a little of the flour-paste mixture. You should have a roll about 7.5 cm long, a little like a small sausage. Repeat the procedure until you have used up all the filling.

Heat the oil in a deep-fat fryer or a large wok until it is hot. Deep-fry the spring rolls, a few at a time, until they are golden brown. They have a tendency to stick to each other at the beginning of the frying, so fry only a few at a time. Do not attempt to break them apart should they stick together. You can do this after removing them from the oil. Drain them on kitchen paper. Serve at once.

Asian Duck Confit

Serves 6

6 fresh or frozen duck thigh and leg (1.5 kg / 3¼ lb)
50 g / 1¾ oz coarse sea salt
4 x 350 g / 12½ oz tins duck or goose fat
8 whole unpeeled garlic cloves, lightly crushed
8 slices fresh ginger
6 whole star anise
3 cinnamon bark or stick
2 tablespoons whole roasted Sichuan peppercorns

Lay the duck on a tray and evenly sprinkle the salt on both sides. Cover the duck with a linen cloth and keep in a cool place or the fridge overnight.

The next day, wipe off the salt and pat dry with linen towels. Heat the duck or goose fat in a large pot, and add the garlic, ginger, star anise, cinnamon and peppercorns. Now add the duck thighs and cook slowly over low heat for 1 hour. The duck should be cooked and tender. Let the duck cool in the fat and refrigerate with the fat covering the duck until ready to use. The confit will keep for months in the fat in the fridge.

When you are ready to cook the duck confit, preheat the oven to 180 °C / gas mark 4.

Scoop the duck from the fat and roast on a tray for 40 minutes or until the duck is crispy. Remove the duck from the hot fat and drain on kitchen paper. Strain the hot fat and keep for future use. Serve at once. You can use the fat to sauté potatoes.

This wonton soup dish is simple and quick to make. It highlights the use of fresh truffles in an unusual vegetarian dish, if you use vegetarian stock. I use wonton skins, which can be obtained from Chinese grocers as well as supermarkets now. They are yellowish in colour, square and packaged in small stacks. They can be bought fresh or frozen but be sure to thaw them thoroughly if they are frozen.

Chinese Wonton Soup with Truffles

Serves 4

175 g / 6 oz, about 1 small packet wonton skins

80 g / 2¾ oz Fresh black winter truffles, sliced

Salt and freshly ground black pepper

1 egg, beaten

1.2 litres / 2 pints homemade vegetarian stock or chicken stock

2 tablespoons fresh chives, finely chopped

1 tablespoon fresh chervil, finely chopped

When you are ready to stuff the wontons, brush one sheet with the beaten egg. Add two to three slices of fresh truffles, and salt and pepper to taste. Then lay another wonton sheet over and pinch the edges closed. Press with your palm to push out any air to seal the wonton.

When the wontons are ready, bring the stock to a simmer in a large pot.

In another large pot, bring salted water to a boil and poach the wontons for 1 minute or until they float to the top. Remove them immediately and transfer them to the pot with the stock. This procedure will result in a cleaner-tasting broth. Continue to simmer them in the stock for 2 minutes. Transfer to either a large soup terrine or individual bowls. Garnish with the fresh herbs and serve immediately.

Steaming is an often-used Chinese cooking method that uses warm vapours to coax flavours from food rather than using dry heat and fat. This subtle technique works extremely well for foie gras and truffles. This is a perfect recipe for a delicious, easy-to-make and impressive first course.

Steamed Foie Gras in Chinese Cabbage with Truffles

Serves 4

4 pieces fresh foie gras (100 g / 3½ oz each)

Salt and freshly ground black pepper to taste

4 large Chinese cabbage leaves (Peking cabbage)

50 g / 1¾ oz fresh winter truffles or highest-quality tinned winter truffles (cooked once), thinly sliced

Generously season each piece of foie gras with salt and pepper.

Blanch the cabbage leaves in a pot of salted water for a few seconds and cool thoroughly.

Place slices of truffles in the centre of each cabbage leaf, then a piece of foie gras, finally finish with more slices of truffles. Roll up each leaf so that the foie gras is covered.

Steam on a plate over simmering water for 5–7 minutes, depending on thickness and size.

When the foie gras is cooked, pour off the liquid and place each foie gras on an individual plate and serve at once.

This Japanese-inspired recipe works wonderfully with truffles. The crispy, lacy batter provides a crackling texture encasing the aromatic truffle. It is essential that you use the highest quality fresh truffles, hot oil at a constant temperature and quite lumpy batter that is well chilled. Make the batter at the very last moment, as it should not sit. You will find this a delicious, elegant opening for any meal.

Crispy Truffle Tempura

Serves 4

100 g / 3½ oz fresh black winter truffles
400 ml / 14 fl oz groundnut or peanut oil
flour, for dusting

Batter:
2 egg yolks
500 ml / 18 fl oz ice water
300 g / 10½ oz flour, sifted
a few ice cubes
fleur de sel or coarse sea salt

Cut the truffles into 5 mm-thick rounds.

Heat the oil in a deep pan until it is hot.

Dust each truffle round with flour, shaking off any excess.

Now make the batter in two batches. First, while the oil is heating, beat one egg yolk, then add half the water (250 ml / 9 fl oz). Finally add half the flour (150 g / 5¼ oz) all at once. Beat a few times to mix lightly; a chopstick is perfect for this. The ingredients should be loosely combined.

Add a couple of ice cubes to the batter to keep it cold.

Dip the truffles into the batter and then drop them in the hot oil. Remove them at once when they are crispy and drain well on kitchen paper. Continue to do the same until you run out of batter. Then make another batch using the ingredients that are remaining.

Arrange on a warm platter, sprinkle with fleur de sel or coarse sea salt and serve at once.

PEBEYRE

ACKNOWLEDGEMENTS

For Monique & Jacques Pébeyre

We are grateful to Jacques and Monique Pébeyre for their support from the conceptual beginning of this book. We wanted to capture their passionate love of truffles through history and through food. We are fortunate to have Shane Suvikapakornkul, who immediately realised the story to be told. Of course, Jean-Pierre Gabriel's magnificent photos illustrated in vivid colours the essence of truffles. Toby Matthews's hard work translating and editing the text and recipes made for a better book. We would have been lost without the help of Babé Pébeyre, Mic Cheminal and Daniel Taurines throughout the testing of the recipes of this book.

Ken Hom and Pierre-Jean Pébeyre

First published in 2014 by Serindia Contemporary,
an imprint of Serindia Publications

Serindia Publications, Inc.
PO Box 10335
Chicago, Illinois 60610 USA
www.serindia.com

ISBN: 978-1932476-68-2 (Hardcover)

A version of the text and recipes appeared in German, Dutch, Flemish and French editions in 2000 and 2001. For this volume, the original French text was completely revised in 2014 by Pierre-Jean Pébeyre and the recipes updated by Ken Hom and Pierre-Jean Pébeyre, with all-new photographs by Jean-Pierre Gabriel.

Translated from revised French texts by Toby Matthews

Designed by Toby Matthews

The Cataloguing-In-Publication record for this book is available from
the Library of Congress
Library of Congress Control Number: 2014950097

Printed in China